ANIMAL TRACKS
of
NEW ENGLAND

Ian Sheldon, Tamara Hartson
&
Mark Elbroch

LONE
PINE

© 2000 by Lone Pine Publishing
First printed in 2000   10 9 8 7 6 5 4 3 2 1
Printed in Canada

THE PUBLISHER: LONE PINE PUBLISHING

1901 Raymond Avenue SW, Suite C	10145-81 Avenue
Renton, WA 98055	Edmonton, AB T6E 1W9
USA	Canada

Lone Pine Publishing website: http://www.lonepinepublishing.com

Canadian Cataloguing in Publication Data

Sheldon, Ian, (date)
  Animal tracks of New England

  Includes bibliographical references and index.
  ISBN 1-55105-246-6

  1. Animal tracks—New England—Identification.
I. Hartson, Tamara, (date) II. Elbroch, Mark, (date) III. Title.
QL768.S5238 1999   591.47'9   C99-911100-0

Editorial Director: Nancy Foulds
Editor: Volker Bodegom
Production Manager: Jody Reekie
Design, layout and production: Volker Bodegom, Monica Triska
Cartography: Volker Bodegom
Animal illustrations: Gary Ross, Ian Sheldon, Kindrie Grove
Track illustrations: Ian Sheldon
Cover illustration: Raccoon by Gary Ross
Scanning: Elite Lithographers Ltd.

We acknowledge the financial support of the Government of Canada
through the Book Publishing Industry Development Program (BPIDP) for
our publishing activities.

PC: P4

CONTENTS

INTRODUCTION

If you have ever spent time with an experienced tracker, or perhaps a veteran hunter, then you know just how much there is to learn about the subject of tracking and just how exciting the challenge of tracking animals can be. Maybe you think that tracking is no fun, because all you get to see is the animal's prints. What about the animal itself—is that not much more exciting? Well, for most of us who don't spend a great deal of time in the beautiful wilderness of New England, the chances of seeing the majestic Moose or the fun-loving River Otter are slim. The closest that we may ever get to some animals will be through their tracks, and they can inspire a very intimate experience. Remember, you are following in the footsteps of the unseen—animals that are in pursuit of prey, or perhaps being pursued as prey.

This book offers an introduction to the complex world of tracking animals. Sometimes tracking is easy. At other times it is an incredible challenge that leaves you wondering just what animal made those unusual tracks. Take this book into the field with you, and it can provide some help with the first steps to identification. Animals tracks and trails are this book's focus; you will learn to recognize subtle differences for both. There are, of course, many additional signs to consider, such as scat and food caches, all of which help you to understand the animal that you are tracking.

Remember, it takes many years to become an expert tracker. Tracking is one of those skills that grows with you as you acquire new knowledge in new situations. Most importantly, you will have an intimate experience with nature. You will learn the secrets of the seldom seen. The more you discover, the more you will want to know and, by developing a good understanding of tracking, you will gain an excellent appreciation of the intricacies and delights of our marvelous natural world.

How to Use This Book

Most importantly, take this book into the field with you! Relying on your memory is not an adequate way to identify tracks. Track identification has to be done in the field or with detailed sketches and notes that you can take

Long-tailed Weasel

home. Much of the process of identification involves circumstantial evidence, so you will have much more success when standing beside the track.

This book is laid out in an easy-to-use format. There is a quick reference appendix to the tracks of all the animals illustrated in the book beginning on p. 124. This appendix is a fast way to familiarize yourself with certain tracks and the content of the book, and it guides you to the more informative descriptions of each animal and its tracks.

Each animal's description is illustrated with the appropriate footprints and the track patterns that it usually leaves. Although these illustrations are not exhaustive, they do show the tracks or groups of prints that you will most likely see. You will find a list of dimensions for the tracks, giving the general range, but there will always be extremes, just as there are with people who have unusually small or large feet. Under the category 'Size' (of animal), the 'greater-than' sign (>) is used when the size difference between the sexes is pronounced.

If you think that you may have identified a track, check the 'Similar Species' section for that animal. This section is designed to help you confirm your conclusions by pointing out other animals that leave similar tracks and showing you ways to distinguish among them.

As you read this book, you will notice an abundance of words such as 'often,' 'mostly' and 'usually.' Unfortunately, tracking will never be an exact science; we cannot expect animals to conform to our expectations, so be prepared for the unpredictable.

0   100 mi
0   100 km

N

CANADA

Quebec

N.B.

St. John River

Maine

Lake
Champlain

MONTPELIER

VT.

AUGUSTA

MOUNTAINS

APPALACHIAN

N.H.

CONCORD

Gulf of Maine

N.Y.

Connecticut
River

Mass.

BOSTON

Cape Cod

HARTFORD

PROVIDENCE

Conn.

R.I.

Atlantic Ocean

N.Y. Long Island

Tips on Tracking

As you flip through this guide, you will notice clear, well-formed prints. Do not be deceived! It is a rare track that will ever show so clearly. For a good, clear print, the perfect conditions are slightly wet, shallow snow that isn't melting, or slightly soft mud that isn't actually wet. These conditions can be rare—most often you will be dealing with incomplete or faint prints, where you cannot even really be sure of the number of toes.

Should you find yourself looking at a clear print, then the job of identification is much easier. There are a number of key features to look for: measure the length and width of the print, count the number of toes, check for claw marks and note how far away they are from the body of the print, and look for a heel. Keep in mind more subtle features, such as the spacing between the toes, whether or not they are parallel, and whether fur on the sole of the foot has made the print less clear.

When you are faced with the challenge of identifying an unclear print—or even if you think that you have made a successful identification from one print alone—look beyond the single footprint and search out others. Do not rely on the dimensions of one print alone, but collect measurements from several prints to get an average impression. Even the prints within one trail can show a lot of variation.

Try to determine which is the fore print and which is the hind, and remember that many animals are built very differently from humans, having larger forefeet than hind feet. Sometimes the prints will overlap, or they can be directly on top of one another in a direct register. For some animals, the fore and hind prints are pretty much the same.

Check out the pattern that the tracks make together in the trail, and follow the trail for as many paces as is necessary for you to become familiar with the pattern. Patterns are very important, and they can be the distinguishing feature between different animals with otherwise similar tracks.

Meadow Jumping Mouse

Follow the trail for some distance, because it may give you some vital clues. For example, the trail may lead you to a tree, indicating that the animal is a climber—or it may lead down into a burrow. This part of tracking can be the most rewarding, because you are following the life of the animal as it hunts, runs, walks, jumps, feeds or tries to escape a predator.

Take into consideration the habitat. Sometimes very similar species can be distinguished by their habitats only—one might be found on the riverbank, whereas another might be encountered just in the dense forest.

Think about your geographical location, too, because some animals have a limited range. This consideration can rule out some species and help you with your identification.

Remember that every animal will at some point leave a print or trail that looks just like the print or trail of a completely different animal!

Finally, keep in mind that if you track quietly, you might catch up with the maker of the prints.

Terms & Measurements

Some of the terms used in tracking can be rather confusing, and they often depend on personal interpretation. For example, what comes to your mind if you see the word 'hopping'? Perhaps you see a person hopping about on one leg—or perhaps you see a rabbit hopping through the countryside. Clearly, one person's perception of motion can be very different from another's. Some useful terms are explained below, to clarify what is meant in this book and, where appropriate, how the measurements given fit in with each term.

The following terms are sometimes used loosely and interchangeably—for example, a rabbit might be described as 'a hopper' and a squirrel as 'a bounder,' yet both leave the same pattern of prints in the same sequence.

Ambling: Fast, rolling walking.

Bounding: A gait of four-legged animals in which the two hind feet land simultaneously, usually registering in front of the forefeet. Common in rodents and the rabbit family. 'Hopping' or 'jumping' can often be substituted.

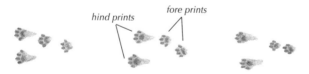

hind prints   *fore prints*

Gait: An animal's gait describes how it is moving at some point in time, and it results in observable trail characteristics.

Galloping: A gait used by animals with four even-length legs, such as dogs, moving at high speed, hind feet registering in front of forefeet.

hind prints   *fore prints*

gallop group

Hopping: Similar to bounding. With four-legged animals, usually indicated by tight clusters of prints, fore prints set between and behind the hind prints. A bird hopping on two feet creates a series of paired tracks along its trail.

Loping: Like galloping but slower, with each foot falling independently and leaving a trail pattern that consists of groups of tracks in the sequence fore-hind-fore-hind, usually roughly in a line.

Mustelids (weasel family) often use ***2x2 loping***, in which the hind feet register directly on the fore prints. The resulting pattern has angled, paired tracks.

Running: Like galloping, but applied generally to animals moving at high speed. Also used for two-legged animals.

Trotting: Faster than walking, slower than running. The diagonally opposite limbs move simultaneously; that is, the right forefoot with the left hind, then the left forefoot with the right hind. This gait is the natural one for canids (dog family), short-tailed shrews and voles.

hind print   fore print

Canids may use ***side-trotting***, a fast trotting in which the hind end of the animal shifts to one side. The resulting track pattern has paired tracks, with all the fore prints on one side and all the hind prints on the other.

Walking: A slow gait in which each foot moves independently of the others, resulting in an alternating track pattern. This gait is common for felines (cat family) and deer, as well as wide-bodied animals, such as bears and porcupines. The term is also used for two-legged animals.

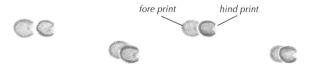

fore print   hind print

Other Tracking Terms:

Dewclaws: Two small, toe-like structures set above and behind the main foot of most hoofed animals.

dewclaw marks

dragline

Direct Register: The hind foot falls directly on the fore print.

double register

direct register

Double Register: The hind foot overlaps the fore print only slightly or falls beside it, so that both prints can be seen at least in part.

Dragline: A line left in snow or mud by a foot or the tail dragging over the surface.

Gallop Group: A track pattern of four prints made at a gallop, usually with hind feet registering in front of forefeet.

fore prints

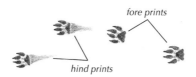

hind prints

Height: Taken at the animal's shoulder.

Length: The animal's body length from head to rump, not including the tail, unless otherwise indicated.

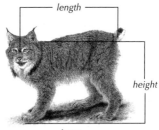

Lynx

Metacarpal Pad: A small pad near the palm pad or between the palm pad and heel on the forefeet of bears and members of the weasel family.

Print (also called '**track**'): Fore and hind prints are treated individually. Print dimensions given are 'length' (including claws—maximum values may represent occasional heel register for some animals) and 'width.' A group of prints made by each of the animal's feet makes up a track pattern.

Register: To leave a mark—said about a foot, claw or other part of an animal's body.

Retractable: Describes claws that can be pulled in to keep them sharp, as with the cat family; these claws do not register in the prints. Foxes have semi-retractable claws.

Sitzmark: The mark left on the ground by an animal falling or jumping from a tree.

Straddle: The total width of the trail, all prints considered.

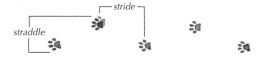

Stride: For consistency among different animals, the stride is taken as the distance from the center of one print (or print group) to the center of the next one. Some books may use the term 'pace.'

Track: Same as '***print***.'

Track Pattern: The pattern left after each foot registers once; a set of prints, such as a gallop group.

Trail: A series of track patterns; think of it as the path of the animal.

Redback Salamander

MAMMALS

River Otter

Moose

Fore and Hind Prints
Length: 4–7 in (10–18 cm)
Length with dewclaws: to 11 in (28 cm)
Width: 3.5–6 in (9–15 cm)

Straddle
8.5–20 in (22–50 cm)

Stride
Walking: 1.5–3 ft (45–90 cm)
Trotting: to 4 ft (1.2 m)

Size (bull>cow)
Height: 5–6.5 ft (1.5–2 m)
Length: 7–8.5 ft (2.1–2.6 m)

Weight
600–1100 lb (270–500 kg)

walking

MOOSE
Alces alces

The impressive male Moose, largest of the deer, has a massive rack of antlers. Moose are usually solitary, though you may see a cow with her calf. Despite its placid appearance, a moose may charge humans if approached.

The ungainly shaped Moose moves gracefully, leaving a neat alternating walking pattern. The hind feet direct or double register on the fore prints. Long legs allow for easy movement in snow; where the print is deeper than 1.2 inches (3 cm), the dewclaws—which give extra support for the animal's great weight—register but far behind the hoof. In summer, look for tracks in mud beside ponds and other wet areas, where Moose especially like to feed; they are excellent swimmers. In winter, Moose feed in willow flats and coniferous forests, leaving a distinct browseline (highline). Ripped stems and scraped bark, 6 feet (1.8 m) or more above the ground, are additional signs of Moose.

Similar Species: White-tailed Deer (p. 20) tracks—smaller, with a narrower straddle—may be mistaken for a juvenile Moose's.

White–tailed Deer

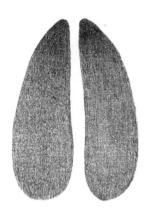

Fore and Hind Prints
Length: 2–3.5 in (5–9 cm)
Width: 1.6–2.5 in (4–6.5 cm)

Straddle
5–10 in (13–25 cm)

Stride
Walking: 10–20 in (25–50 cm)
Galloping: 6–15 ft (1.8–4.5 m)

Size (buck>doe)
Height: 3–3.5 ft (90–110 cm)
Length: to 6.3 ft (1.9 m)

Weight
120–350 lb (55–160 kg)

walking   *gallop group*

WHITE-TAILED DEER
Odocoileus virginianus

The keen hearing of this deer guarantees that it knows about you before you know about it. Frequently, all that we see is its conspicuous white tail in the distance as it gallops away, earning this deer the nickname 'flagtail.' This adaptable deer may be found throughout New England, in small groups at the edges of forests and in brushlands. The White-tailed Deer can be common around ranches and residential areas.

This deer's prints are heart-shaped and pointed. Its alternating walking track pattern shows the hind prints direct or double registered on the fore prints. In snow or when a deer gallops on soft surfaces, the dewclaws register. This flighty deer gallops in the usual style, hind prints ahead of fore prints, with toes spread wide for better footing.

Similar Species: Juvenile Moose (p. 18) tracks may be confused with large deer tracks.

Horse

Fore Print
(hind print is slightly smaller)
Length: 4.5–6 in (11–15 cm)
Width: 4.5–5.5 in (11–14 cm)

Stride
Walking: 17–28 in (43–70 cm)

Size
Height: to 6 ft (1.8 m)

Weight
to 1500 lb (680 kg)

walking

HORSE
Equus caballus

Back-country use of the popular Horse means that you can expect its tracks to show up almost anywhere.

Unlike any other animal in this book, the Horse has only one huge toe. This toe leaves an oval print with a distinctive 'frog' (V-shaped mark) at its base. When the Horse is shod, the horseshoe shows up clearly as a firm wall at the outside of the print. Not all horses are shod, so do not expect to see this outer wall on every horse print. A typical, unhurried horse trail is an alternating walking pattern, with the hind prints registered on or behind the slightly larger fore prints. Horses are capable of a range of speeds—up to a full gallop—but most recreational horse-back riders take a more leisurely outlook on life, preferring to walk their horses and soak up the mountain views!

Similar Species: Mules (rarely shod) have smaller tracks.

Black Bear

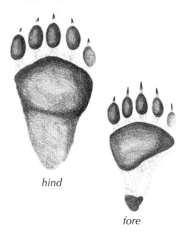

hind

fore

Fore Print
Length: 4–6.3 in (10–16 cm)
Width: 3.8–5.5 in (9.5–14 cm)

Hind Print
Length: 6–7 in (15–18 cm)
Width: 3.5–5.5 in (9–14 cm)

Straddle
9–15 in (23–38 cm)

Stride
Walking: 17–23 in (43–58 cm)

Size (male>female)
Height: 3–3.5 ft (90–110 cm)
Length: 5–6 ft (1.5–1.8 m)

Weight
200–600 lb (90–270 kg)

walking (slow)

BLACK BEAR
Ursus americanus

The Black Bear has a scattered distribution throughout New England's forested areas, but do not expect to see its tracks in winter, when it sleeps deeply. Finding fresh bear tracks can be a thrill, but take care—the bear may be just ahead. Never underestimate the potential power of a surprised bear!

Black Bear prints resemble small human prints, but they are wider with claw marks. The small inner toe rarely registers. The forefoot's small heel pad often shows, and the hind foot has a big heel. The bear's slow walk results in a slightly pigeon-toed double register with the hind print on the fore print. More frequently, at a faster pace, the hind foot oversteps the forefoot. When a bear runs, the two hind feet register in front of the forefeet in an extended cluster. Along well-worn bear paths, look for 'digs'—patches of dug-up earth—and 'bear trees' whose scratched bark shows that these bears climb.

Similar Species: The magnificent Grizzly (Brown) Bear (*Ursus arctos*) has similar prints but no longer lives here.

Coyote

fore

hind

Fore Print
(hind print is slightly smaller)
Length: 2.4–3.2 in (6–8 cm)
Width: 1.6–2.4 in (4–6 cm)

Straddle
4–7 in (10–18 cm)

Stride
Trotting: 17–23 in (43–58 cm)
Galloping: 2.5–10 ft (0.8–3 m)

Size (female is slightly smaller)
Height: 23–26 in (58–65 cm)
Length: 32–40 in (80–100 cm)

Weight
20–50 lb (9–23 kg)

trotting   *gallop group*

COYOTE
(Brush Wolf, Prairie Wolf)
Canis latrans

This widespread, adaptable canine prefers open grasslands or woodlands. It hunts rodents and larger prey, on its own, with a mate or in a family pack. If you find a Coyote den—usually a wide-mouthed tunnel leading into a nesting chamber—take care not to bother the family or the female will have to move her pups to a safer location.

The oval fore prints are slightly larger than the hind prints. The fore heel pad is more triangular than the hind heel pad, which rarely registers clearly. The two outer toe prints usually lack claw marks. A Coyote's tail hangs down, leaving a dragline in deep snow. A Coyote typically walks or trots in an alternating pattern; the walk has a wider straddle. A Coyote's trotting trail is often very straight. When a Coyote gallops, its hind feet fall in front of its fore-feet; the faster it goes, the straighter the gallop group.

Similar Species: A Domestic Dog's (*Canis familiaris*) less-oval prints splay more, and its trail is erratic. Foot hairs make Red Fox (p. 28) prints (usually smaller) less clear. Gray Fox (p. 30) tracks are much smaller.

Red Fox

fore

hind

**Fore Print
(hind print is slightly smaller)**
Length: 2.1–3 in (5.3–7.5 cm)
Width: 1.6–2.3 in (4–5.8 cm)

Straddle
2–3.5 in (5–9 cm)

Stride
Trotting: 12–18 in (30–45 cm)
Side-trotting:
14–21 in (36–53 cm)

Size (vixen is slightly smaller)
Height: 14 in (35 cm)
Length: 22–25 in (55–65 cm)

Weight
7–15 lb (3.2–7 kg)

trotting | side-trotting

28

RED FOX

Vulpes vulpes

This beautiful and notoriously cunning fox, found throughout the region, prefers mountainous forests and open areas. It is a very adaptable and intelligent animal.

The finer details of a Red Fox's tracks are obscured by its foot hairs; only parts of the toes and heel pads show. The horizontal or slightly curved bar across the fore heel pad is diagnostic. A Red Fox leaves a distinctive, straight trail of alternating prints—the hind print direct registers on the wider fore print. When a fox side-trots, it leaves print pairs with the hind print falling to one side of the fore print in typical canid fashion. Foxes gallop like Coyotes (p. 26). The faster the gallop, the straighter the gallop group.

Similar Species: Other canid prints lack the bar across the fore heel pad. Gray Fox (p. 30) prints are smaller. Domestic Dog (*Canis familiaris*) prints can be of similar size. Small Coyote prints are similar but have a wider straddle and more bulbous toe registrations.

Gray Fox

fore

hind

**Fore Print
(hind print slightly smaller)**
Length: 1.3–1.8 in (3.3–4.5 cm)
Width: 1.1–1.5 in (2.8–3.8 cm)

Straddle
2–4 in (5–10 cm)

Stride
Trotting: 10–14 in (25–35 cm)

Size
Height: 14 in (35 cm)
Length: 21–30 in (53–75 cm)

Weight
7–15 lb (3.2–7 kg)

trotting

GRAY FOX

Urocyon cinereoargenteus

This small, shy fox is widespread, but it especially prefers woodlands and chaparral country. This fox is the only one that climbs trees, which it does either for safety or to forage.

The larger forefoot registers better than the hind foot, for which the long, semi-retractable claws do not always register. The heel pads are often unclear—they sometimes show up just as small, round dots. This fox has a neat alternating trot pattern. Its gallop group is like the Coyote's (p. 26).

Similar Species: The Red Fox (p. 28) has heel pads with a bar across them; in general, its prints are larger and less clear (because of thick fur), its stride is longer, and its straddle is narrower. Coyote tracks are much larger. Domestic Cat (p. 38) and Bobcat (p. 36) prints lack claw marks and have larger, less symmetrical heel pads.

Mountain Lion

fore

hind

Fore Print
(hind print is slightly smaller)
Length: 3–4.5 in (7.5–11 cm)
Width: 3.3–4.8 in (8.5–12 cm)

Straddle
8–12 in (20–30 cm)

Stride
Walking: 13–32 in (33–80 cm)
Bounding: to 12 ft (3.7 m)

Size
Height: 26–32 in (65–80 cm)
Length: 3.5–5 ft (1.1–1.5 m)

Weight
70–200 lb (32–90 kg)

walking (fast)

MOUNTAIN LION
(Cougar, Puma, Panther)

Puma concolor

This elusive animal is extremely rare in New England, and the few sightings may be escaped or intentionally released animals. If you have conclusive evidence of a Mountain Lion, please notify the U.S. Fish and Wildlife Service—you have made a noteworthy discovery.

Mountain Lion prints tend to be wider than long; the retractable claws never register. In winter, thick fur makes the prints much larger, and it may obscure the two lobes on the front of the heel pad. When a Mountain Lion walks, the hind foot either direct or double registers on the larger fore print—the hind feet overstep only if the animal is walking fast. In snow, the long, thick tail may drag and obscure the prints. When chasing prey, the Mountain Lion makes long bounds.

Similar Species: The Lynx (p. 34) has prints similar in size, but it has a narrower straddle, a shorter stride and no tail drag, and it does not sink as deep in snow. Bobcat (p. 36) prints may be confused with juvenile Mountain Lion prints but are smaller and clearer and sink deeper in snow.

Lynx

fore

hind

Fore Print
(hind print is slightly smaller)
Length: 3.5–4.5 in (9–11 cm)
Width: 3.5–4.8 in (9–12 cm)

Straddle
6–9 in (15–23 cm)

Stride
Walking: 12–28 in (30–70 cm)

Size
Length: 2.5–3 ft (75–90 cm)

Weight
15–30 lb (7–14 kg)

walking

LYNX
Lynx canadensis

This large cat is a thrill to see, but it is sensitive to human interference. In New England, the elusive Lynx inhabits only the remote, undisturbed, dense northern forests. With huge feet and a relatively lightweight body, it stays on top of the snow as it pursues its main prey, the Snowshoe Hare (p. 60).

This cautious walker leaves a neat alternating track pattern, the hind print direct registered on top of the fore print. Thick fur on the feet often results in prints that are big, round depressions with no detail. In deeper snow, the print may be extended by 'handles' off to the rear. However deep the snow, this cat sinks no more than 8 inches (20 cm), and it rarely drags its feet. A Lynx is more likely to bound than to run. Its curious nature results in a meandering trail that may lead to a partially buried food cache.

Similar Species: Bobcat (p. 36) prints are smaller, and the trail may show draglines. Fisher (p. 48) prints may not show the fifth toe—look for mustelid habits. Coyote (p. 26), Fox (pp. 28–31) and Domestic Dog (*Canis familiaris*) prints show claw marks and their length exceeds their width.

Bobcat

fore

hind

Fore Print
(hind print is slightly smaller)
Length: 1.8–2.5 in (4.5–6.5 cm)
Width: 1.8–2.5 in (4.5–6.5 cm)

Straddle
4–7 in (10–18 cm)

Stride
Walking: 8–16 in (20–40 cm)
Running: 4–8 ft (1.2–2.4 m)

Size
(female is slightly smaller)
Height: 20–22 in (50–55 cm)
Length: 25–30 in (65–75 cm)

Weight
15–35 lb (7–16 kg)

walking

*ambling
to loping*

BOBCAT (Wildcat)

Lynx rufus

The Bobcat, widely distributed in New England, is seldom seen. This stealthy hunter usually hunts by night. Very adaptable, it can leave tracks anywhere from wild mountainsides to chaparral and even into residential areas.

When a Bobcat walks, its hind feet usually register directly on the larger fore prints. As a Bobcat picks up speed, its trail becomes an ambling pattern of paired prints, the hind leading the fore. At even greater speeds, it leaves four-print groups in a lope pattern. Especially the fore prints show asymmetry. The front part of the heel pad has two lobes and the rear part three. In deep snow, the Bobcat's feet leave draglines. The Bobcat marks its territory with half-buried scat along its meandering trail.

Similar Species: A large Domestic Cat (p. 38) has similar prints but has a shorter stride and a narrower straddle. Fisher (p. 48) hind prints that lack the fifth toe may seem similar—look for mustelid habits. Fox (pp. 28–31), Coyote (p. 26) and Domestic Dog (*Canis familiaris*) prints are narrower than long and show claw marks, and the fronts of their footpads are once-lobed; wild canid trails do not meander.

Domestic Cat

fore

hind

Fore Print
(hind print is slightly smaller)
Length: 1–1.6 in (2.5–4 cm)
Width: 1–1.8 in (2.5–4.5 cm)

Straddle
2.4–4.5 in (6–11 cm)

Stride
Walking: 5–8 in (13–20 cm)
Loping/Galloping:
    14–32 in (35–80 cm)

Size (male>female)
Height: 20–22 in (50–55 cm)
Length with tail: 30 in (75 cm)

Weight
6.5–13 lb (3–6 kg)

walking

loping to
galloping

DOMESTIC CAT
(House Cat)
Felis catus

The tracks of the familiar and abundant Domestic Cat can show up almost any place where there are people. Abandoned cats may roam further afield; these 'feral cats' lead a pretty wild and independent existence. Domestic Cats can come in many shapes, sizes and colors.

As with all felines, a Domestic Cat's fore print and slightly smaller hind print both show four toe pads. Its retractable claws, kept clean and sharp for catching prey, do not register. Cat prints usually show a slight asymmetry, with one toe leading the others. A Domestic Cat makes a neat alternating walking track pattern, usually in direct register, as one would expect from this animal's fastidious nature. When a cat picks up speed, it leaves clusters of four prints, the hind feet registering in front of the forefeet.

Similar Species: A small Bobcat (p. 36) may leave tracks similar to a very large Domestic Cat's. Fox (pp. 28–31) and Domestic Dog (*Canis familiaris*) prints show claw marks.

Raccoon

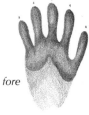

fore

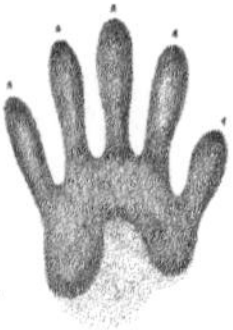

hind

Fore Print
Length: 2–3 in (5–7.5 cm)
Width: 1.8–2.5 in (4.5–6.5 cm)

Hind Print
Length: 2.4–3.8 in (6–9.5 cm)
Width: 2–2.5 in (5–6.5 cm)

Straddle
3.3–6 in (8.5–15 cm)

Stride
Walking: 8–18 in (20–45 cm)
Galloping: 15–25 in (38–65 cm)

Size
(female is slightly smaller)
Length: 24–37 in (60–95 cm)

Weight
11–35 lb (5–16 kg)

walking   *gallop group*

RACCOON
Procyon lotor

The inquisitive Raccoon, common in New England, is often adored for its distinctive face mask, yet disliked for its boundless curiosity—often demonstrated with residential garbage cans. A good place to look for its tracks is near water at low elevations. Raccoons like to rest in trees, and they usually den up for the colder months.

The Raccoon's unusual print, showing five well-formed toes, looks like a human handprint. The small claws appear as dots. Its highly dexterous forefeet rarely leave heel prints, but its hind prints, which are generally much clearer, do show heels. The Raccoon's peculiar walking track pattern shows the left fore print next to the right hind print (or just in front) and vice versa. On the rare occasions when a Raccoon is out in deep snow, it may use a direct-registering walk. Raccoons occasionally run, leaving clusters where the two hind prints fall in front of the fore prints.

Similar Species: Unclear Opossum (p. 42) prints may look similar, but the Opossum drags its tail. In deep snow, Fisher (p. 48) or River Otter (p. 46) tracks may look similar.

Opossum

fore

hind

Fore Print
Length: 2–2.3 in (5–5.8 cm)
Width: 2–2.3 in (5–5.8 cm)
Hind Print
Length: 2.5–3 in (6.5–7.5 cm)
Width: 2–3 in (5–7.5 cm)
Straddle
4–5 in (10–13 cm)
Stride
5–11 in (13–28 cm)
Size
Length: 2–2.5 ft (60–75 cm)
Weight
9–13 lb (4–6 kg)

walking

fast walk

42

OPOSSUM
Didelphis virginiana

This slow-moving, nocturnal marsupial is found throughout New England. Though it occupies many habitats, it prefers open woodland or brushland around waterbodies. It is quite tolerant of residential areas. Opossum tracks can often be seen in mud near the water and in snow during the warmer months of winter—Opossums tend to den up during severe weather.

Opossums are excellent climbers, so do not be surprised if a trail leads to a tree. They have two walking habits: the common alternating pattern, with the hind prints registering on the fore prints, and a Raccoon-like (p. 40) paired-print pattern, with each hind print next to the opposing fore print. The very distinctive, long, inward-pointing thumb of the hind foot does not make a claw mark. In snow, the dragline of the long, naked tail may be stained with blood; unfortunately, this thinly haired animal is not well adapted to the cold and frequently suffers frostbite.

Similar Species: Prints in which the distinctive thumbs do not show, as in sand or fine snow, may be mistaken for a Raccoon's.

Harbor Seal

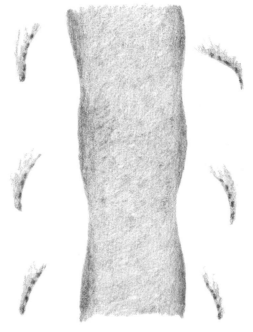

beach track

Size (male>female)
Length: 4–6 ft (1.2–1.8 m)
Weight
180–310 lb (80–140 kg)

HARBOR SEAL

Phoca vitulina

This common seal frequents isolated coastal beaches. One of the smaller seals, it is quite shy and will usually slide off its rocky sentry post into the sea to escape curious naturalists. Look in sandy and muddy areas between rocky platforms for its tracks—unmistakable by their location and large size. Sometimes a Harbor Seal works its way up a river, so you may find its tracks along riverbanks.

Seals, not the most graceful of animals on land, are unrivaled in the water. Their heavy, fat bodies and flipper-like feet can leave messy tracks: a wide trough (made by their cumbersome bellies) with dents alongside (made as the seals pushed themselves along with their forefeet). Look for the marks made by the nails.

Similar Species: The Gray Seal (*Halichoerus grypus*), more than twice as large as the little Harbor Seal, may be found on rocky (sometimes sandy) coasts of New England.

River Otter

fore

hind

Fore Print
Length: 2.5–3.5 in (6.5–9 cm)
Width: 2–3 in (5–7.5 cm)

Hind Print
Length: 3–4 in (7.5–10 cm)
Width: 2.3–3.3 in (5.8–8.5 cm)

Straddle
4–9 in (10–23 cm)

Stride
Loping: 12–27 in (30–70 cm)

Size
(female is two-thirds the size of male)
Length with tail: 3–4.3 ft (90–130 cm)

Weight
10–25 lb (4.5–11 kg)

loping (fast)

RIVER OTTER
Lontra canadensis

No animal knows how to have more fun than a River Otter. If you are lucky enough to watch one at play, you will not soon forget the experience. Widespread and well-adapted for the aquatic environment, this otter lives along waterbodies. Expect to find a wealth of evidence along riverbanks in an otter's home territory. An otter in the forest is usually on its way to another waterbody. Otters love to slide in snow, often down riverbanks, leaving troughs nearly 1 foot (30 cm) wide. In summer they roll and slide on grass and mud.

In soft mud, the River Otter's five-toed feet, especially the hind ones, register evidence of webbing. The inner toes are set slightly apart. If the forefoot's metacarpal pad registers, it lengthens the print. Very variable, otter trails usually show the typical mustelid 2x2 loping, but they show groups of four and three prints with faster gaits. The thick, heavy tail often leaves a dragline.

Similar Species: Mink (p. 52) prints are about half the size. Martens (p. 50), with similar-sized prints, are found in forested regions. Mink and Marten trails lack conspicuous tail drag. Fishers (p. 48) have hairy feet, with the forefeet larger than the hind feet.

47

Fisher

fore

Fore Print
Length: 2.1–4 in (5.3–10 cm)
Width: 2.1–3.3 in (5.3–8.5 cm)

Hind Print
Length: 2.1–3 in (5.3–7.5 cm)
Width: 2–3 in (5–7.5 cm)

Straddle
3–7 in (7.5–18 cm)

Stride
Walking: 7–14 in (18–35 cm)
2x2 loping: 1–4.3 ft (30–130 cm)

Size (male>female)
Length with tail:
    34–40 in (85–100 cm)

Weight
3–12 lb (1.4–5.5 kg)

walking   *2x2 loping*

FISHER (Black Cat)

Martes pennanti

This agile hunter is comfortable both on the ground and in the trees of mixed hardwood forests. The Fisher's speed and eager hunting antics make for exciting tracking as it races up trees and along the ground in its quest for squirrels. It is one of the few predators to kill and eat Porcupines (p. 64).

Though all five toes may register, the small inner toe frequently does not. Only the forefoot has a small heel pad that can show up in the print. The Fisher occasionally walks, making a direct-registering alternating track pattern, but it more often 2x2 lopes in typical mustelid fashion, leaving angled print pairs with the hind print direct registered on the fore print. Loping, its most common gait, produces three- and four- print groups (see the River Otter, p. 46). The patterns often vary within a short distance. Fishers are not associated with water, so 'Fisher' is a misnomer!

Similar Species: Male Marten (p. 50) tracks may be confused with a small female Fisher's, but Martens weigh less and leave shallower prints. Otters have larger hind feet than forefeet. When only four toes register, Fisher prints may look like a Bobcat's (p. 36).

Marten

Fore and Hind Prints
Length: 1.8–2.5 in (4.5–6.5 cm)
Width: 1.5–2.8 in (3.8–7 cm)

Straddle
2.5–4 in (6.5–10 cm)

Stride
Walking: 4–9 in (10–23 cm)
2x2 loping: 9–46 in (23–120 cm)

Size (male>female)
Length with tail:
    21–28 in (53–70 cm)

Weight
1.5–2.8 lb (0.7–1.3 kg)

walking    *2x2 loping*

MARTEN
(American Sable)
Martes americana

This aggressive predator is found in the coniferous and mixed-wood forests of New England.

The Marten seldom leaves a clear print: often just four toes register, the heel pad is undeveloped and, in winter, the hairiness of the feet often obscures all pad detail, especially the poorly developed palm pads. In the Marten's alternating walk, the hind foot registers on the fore print. In 2x2 loping, the hind prints fall on the fore prints to form slightly angled print pairs, in a typical mustelid pattern. Loping track patterns may appear as three- or four-print clusters (see the River Otter, p. 46). Follow the criss-crossing trails—if a Marten has scrambled up a tree, look for a sitzmark where it has jumped down.

Similar Species: Size and habitat are often key to distinguishing Marten, Fisher (p. 48) and Mink (p. 52) tracks. A female Fisher's prints may resemble a large male Marten's, but they will be clearer. Male Mink prints overlap in size with small female Marten prints, but Minks rarely climb trees and, unlike Martens, are usually found near water.

Mink

fore

hind

Fore and Hind Prints
Length: 1.3–2 in (3.3–5 cm)
Width: 1.3–1.8 in (3.3–4.5 cm)

Straddle
2.1–3.5 in (5.3–9 cm)

Stride
Walking/2x2 loping: 8–36 in (20–90 cm)

Size (male>female)
Length with tail: 19–28 in (48–70 cm)

Weight
1.5–3.5 lb (0.7–1.6 kg)

2x2 loping

MINK
Mustela vison

The lustrous Mink, widespread throughout the region, prefers watery habitats surrounded by brush or forest. At home as much on land as in water, this nocturnal hunter can be exciting to track. Like the River Otter (p. 46), the Mink sometimes slides in snow, carving out a trough up to 6 inches (15 cm) wide for an observant tracker to spot.

The Mink's fore print shows five (perhaps four) toes, with five loosely connected palm pads in an arc, but the hind print shows only four palm pads. The metacarpal pad of the forefoot rarely registers, but the furred heel of the hind foot may register, lengthening the hind print. The Mink prefers the typical mustelid 2x2 loping, making consistently spaced, slightly angled double prints. Its diverse track patterns also include alternating walking; loping with three- and four-print groups (like the River Otter); and bounding, like hares (pp. 60–63).

Similar Species: Small Martens (p. 50) may have similar prints, but without a consistent 2x2 loping gait and they do not live near water. Weasels (pp. 54–57) make similar, smaller tracks.

Long-tailed Weasel

Fore and Hind Prints
Length: 1.1–1.8 in (2.8–4.5 cm)
Width: 0.8–1 in (2–2.5 cm)
Straddle
1.8–2.8 in (4.5–7 cm)
Stride
2x2 loping: 9.5–43 in (24–110 cm)
Size (male>female)
Length with tail:
    12–22 in (30–55 cm)
Weight
3–12 oz (85–340 g)

2x2 loping

LONG-TAILED WEASEL

Mustela frenata

The Long-tailed Weasel, the largest of the two weasels in New England, is widely distributed. To identify the weasel species, pay close attention to the straddle, stride and loping patterns, and note the distribution and habitat.

The typical weasel gait is a 2x2 lope, leaving a trail of paired prints. Because of a weasel's light weight and small, hairy feet, the pad detail is often unclear, especially in snow. Even with clear tracks, the inner (fifth) toe rarely registers.

The Long-tailed Weasel's typical 2x2 lope shows an irregularity in the length of the stride—sometimes short and sometimes long—with no consistent behavior. Like the Mink (p. 52), this weasel may bound like a rabbit or hare (pp. 60–63).

Similar Species: A large male Short-tailed Weasel's (p. 56) tracks may be the same size as a small female Long-tailed Weasel's.

Short-tailed Weasel

Fore and Hind Prints
Length: 0.8–1.3 in (2–3.3 cm)
Width: 0.5–0.6 in (1.3–1.5 cm)

Straddle
1–2.1 in (2.5–5.3 cm)

Stride
2x2 loping: 9–36 in (23–90 cm)

Size (male>female)
Length with tail: 8–14 in (20–35 cm)

Weight
1–6 oz (30–170 g)

2x2 loping

SHORT-TAILED WEASEL
(Ermine, Stoat)
Mustela erminea

The Short-tailed Weasel, the smallest of the two weasels in New England, is also widely distributed. It prefers woodlands and meadows up to higher elevations but does not favor wetlands or dense coniferous forests.

Weasels are active year-round hunters, with an avid appetite for rodents. Following their tracks can reveal much about these nimble creatures' activities. Tracks are most evident in winter, when weasels frequently burrow into the snow or pursue rodents into their holes. Some weasel trails may lead you up a tree. Weasels sometimes take to water.

This weasel's 2x2 loping tracks may fall in clusters, with alternating short and long strides.

Similar Species: Small female Long-tailed Weasel (p. 54) tracks may be the same size as a large male Short-tailed's.

Striped Skunk

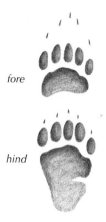

fore

hind

Fore Print
Length: 1.5–2.2 in (3.8–5.6 cm)
Width: 1–1.5 in (2.5–3.8 cm)

Hind Print
Length: 1.5–2.5 in (3.8–6.5 cm)
Width: 1–1.5 in (2.6–3.8 cm)

Straddle
2.8–4.5 in (7–11 cm)

Stride
Walking/Galloping:
    2.5–8 in (6.5–20 cm)

Size
Length with tail:
    20–32 in (50–80 cm)

Weight
6–14 lb (2.7–6.5 kg)

walking (fast)   *galloping*

STRIPED SKUNK
Mephitis mephitis

This striking skunk has a notorious reputation for its vile smell; the lingering odor is often the best sign of its presence. Widespread throughout New England, it prefers lower elevations in a diversity of habitats. The Striped Skunk dens up in winter, coming out on warmer days and in spring.

Both the forefeet and hind feet have five toes. The long claws on the forefeet often register. Smooth palm pads and small heel pads leave surprisingly small prints. A skunk mostly walks—with such a potent smell for its defense, and those memorable black-and-white stripes, it rarely needs to run. Note that this skunk's trail rarely shows any consistent pattern, but an alternating walking pattern may be evident. The greater a skunk's speed, the more the hind foot oversteps the fore. If it runs, its trail consists of clumsy, closely set four-print groups. In snow it drags its feet.

Similar Species: No other New England animal makes comparable tracks. Mustelid (pp. 46–57) tracks are further apart than those resulting from a skunk's shuffling gait. Skunk prints do not overlap.

Snowshoe Hare

hind

fore

Fore Print
Length: 2–3 in (5–7.5 cm)
Width: 1.5–2 in (3.8–5 cm)
Hind Print
Length: 4–6 in (10–15 cm)
Width: 2–3.5 in (5–9 cm)
Straddle
6–8 in (15–20 cm)
Stride
Hopping: 0.8–4.3 ft (25–130 cm)
Size
Length: 12–21 in (30–53 cm)
Weight
2–4 lb (0.9–1.8 kg)

hopping

SNOWSHOE HARE
(Varying Hare)

Lepus americanus

This hare is well known for its color change from summer brown to winter white and for its huge hind feet, which enable it to 'float' on top of snow. Widespread throughout New England, it frequents brushy areas in forests, which provide good cover from the Lynx (p. 34) and the Coyote (p. 26), its most likely predators. Hares are most active at night.

As with other hares and rabbits, the Snowshoe Hare's most common track pattern is a hopping one, with triangular four-print groups; they can be quite long if the hare moves quickly. In winter, heavy fur on the hind feet (much larger than the forefeet) thickens the toes, which can splay out to further distribute the hare's weight on snow. A hare makes well-worn runways that are often used as escape runs. You may encounter a resting hare, because hares do not live in burrows. Twigs and stems neatly cut at a 45° angle also indicate this hare's presence.

Similar Species: New England Cottontail (*Sylvilagus transitionalis*) prints are slightly smaller. European Hare (p. 62) prints are much larger.

European Hare

fore

hind

Fore Print
Length: 2–4 in (5–10 cm)
Width: 1.3 in (3.3 cm)

Hind Print
Length: 5 in (13 cm)
Length with heel: 9 in (23 cm)
Width: 2.5 in (6.5 cm)

Straddle
8 in (20 cm)

Stride
Hopping: 1.5–3 ft (45–90 cm)
Bounding: to 12 ft (3.7 m)

Size
Length: 25–30 in (65–75 cm)

Weight
7–12 lb (3.2–5.5 kg)

hopping

EUROPEAN HARE
Lepus europaeus

This large, sturdy colonizer of open fields was introduced to North America over one hundred years ago. It can now be found throughout southern New England. In winter the European Hare remains brown—unlike the Snowshoe Hare (p. 60), which turns white—and it has a black spot on its tail.

The usual track of this hare is typical for rabbits and hares: elongated triangular clusters of prints. The forefeet register first (usually in a slightly diagonal line), leaving small, roundish prints that show four toes. The two much larger hind prints then fall (usually side by side) ahead of the fore prints. The hind prints are much longer when the whole heel registers, as when the hare stops for a moment.

Similar Species: The much smaller New England Cottontail (*Sylvilagus transitionalis*) makes smaller tracks, with a narrower straddle and smaller stride. The hind toes of the much smaller Snowshoe Hare splay out more in the snow.

Porcupine

fore

hind

Fore Print
Length: 2.3–3.3 in (5.8–8.5 cm)
Width: 1.3–1.9 in (3.3–4.8 cm)
Hind Print
Length: 2.8–4 in (7–10 cm)
Width: 1.5–2 in (3.8–5 cm)
Straddle
5.5–9 in (14–23 cm)
Stride
Walking: 5–10 in (13–25 cm)
Size
Length with tail: 25–40 in (65–100 cm)
Weight
10–28 lb (4.5–13 kg)

walking

PORCUPINE

Erethizon dorsatum

This notorious rodent rarely runs—its many long quills are a formidable defense. Widespread in this region, the Porcupine shows a preference for forests but can also be seen in more open areas.

The Porcupine's preferred pigeon-toed, waddling gait leaves an alternating walking pattern, with the hind print registered on or slightly in front of the shorter fore print. Look for long claw marks on all prints. The fore print shows four toes, and the hind print shows five. On clear prints, the unusual pebbly surface of the solid heel pads may show, but a Porcupine's tracks are often obscured by scratches from its heavy, spiny tail. In deeper snow this squat animal drags its feet, and it may leave a trough with its body. A Porcupine's trail might lead you to a tree, where this animal spends much of its time feeding—look for chewed bark or nipped twigs on the ground.

Similar Species: The Porcupine's distinctive prints differ from those of any other animal of similar size in this region.

Beaver

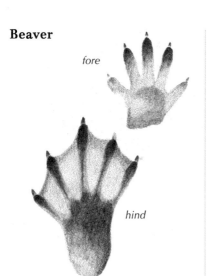

fore

hind

Fore Print
Length: 2.5–4 in (6.5–10 cm)
Width: 2–3.5 in (5–9 cm)

Hind Print
Length: 5–7 in (13–18 cm)
Width: 3.3–5.3 in (8.5–13 cm)

Straddle
6–11 in (15–28 cm)

Stride
Walking: 3–6.5 in (7.5–17 cm)

Size
Length with tail:
    3–4 ft (90–120 cm)

Weight
28–75 lb (13–34 kg)

walking

BEAVER
Castor canadensis

Few animals leave as many signs of their presence as the Beaver, the largest North American rodent and a common sight around water. Look for the conspicuous dams and lodges—capable of changing the local landscape—and the stumps of felled trees. Check trunks gnawed clean of bark for marks of the Beaver's huge incisors. A scent mound marked with castoreum, a strong-smelling yellowish fluid that Beavers produce, also indicates recent activity.

The thick, scaly tail of the Beaver may mar its tracks, as can the branches that the Beaver drags about for construction and food. Check the large hind prints for signs of webbing and broad toenails—the nail of the second inner toe usually does not show. Rarely do all five toes on each foot register. Irregular foot placement in the alternating walking gait may produce a direct or a double register. Repeated path use results in well-worn trails.

Similar Species: The Beaver's many signs, including large hind prints, minimize confusion. Muskrat (p. 68) prints are smaller.

Muskrat

fore

hind

Fore Print
Length: 1.1–1.5 in (2.8–3.8 cm)
Width: 1.1–1.5 in (2.8–3.8 cm)

Hind Print
Length: 1.6–3.2 in (4–8 cm)
Width: 1.5–2.1 in (3.8–5.3 cm)

Straddle
3–5 in (7.5–13 cm)

Stride
Walking: 3–5 in (7.5–13 cm)
Running: to 1 ft (30 cm)

Size
Length with tail: 16–25 in (40–65 cm)

Weight
2–4 lb (0.9–1.8 kg)

walking

MUSKRAT
Ondatra zibethicus

Like Beavers (p. 66), these rodents are found throughout New England, wherever there is water. Beavers are very tolerant of Muskrats and even allow them to live in parts of their lodges. Active all year, Muskrats leave plenty of signs. They dig extensive networks of burrows, often undermining riverbanks, so do not be surprised if you suddenly fall into a hidden hole! Also look for small lodges in the water and beds of vegetation on which Muskrats rest, sun and feed in summer.

The small inner toe of the five on each forefoot rarely registers. The hind print shows five well-formed toes that may have a 'shelf' around them, created by stiff hairs that aid in swimming. The common alternating walking pattern shows track pairs that alternate from side to side, with the hind print just behind or slightly overlapping the fore print. In snow, a Muskrat's feet drag and its tail leaves a sweeping dragline.

Similar Species: Few animals share this water-loving rodent's habits. Beavers make larger prints and leave many other signs.

69

Woodchuck

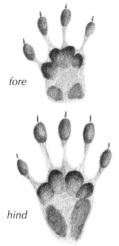

fore

hind

Fore and Hind Prints
Length: 1.8–2.8 in (4.5–7 cm)
Width: 1–2 in (2.5–5 cm)

Straddle
3.3–6 in (8.5–15 cm)

Stride
Walking: 2–6 in (5–15 cm)
Bounding: 6–14 in (15–35 cm)

Size (male>female)
Length with tail:
    20–25 in (50–65 cm)

Weight
5.5–12 lb (2.5–5.5 kg)

walking   *bounding*

WOODCHUCK
(Whistle Pig, Groundhog, Marmot)

Marmota monax

This robust member of the squirrel family is a common sight in open woodlands and adjacent open areas throughout New England. Always on the watch for predators, but not too troubled by humans, the Woodchuck never wanders far from its burrow. This marmot hibernates during winter but emerges in early spring; look for its tracks in late spring snowfalls and in the mud around the burrow entrances.

A Woodchuck's fore print shows four toes, three palm pads and two heel pads, which are not always evident. The hind print shows five toes, four palm pads and two poorly registering heel pads. The Woodchuck usually leaves an alternating walking pattern, with the hind print registered on the fore print. When a Woodchuck runs from danger, it makes groups of four prints, hind ahead of fore.

Similar Species: Small Raccoon (p. 40) running tracks are similar, but they show five-toed fore prints.

Eastern Chipmunk

fore

hind

Fore Print
Length: 0.8–1 in (2–2.5 cm)
Width: 0.4–0.8 in (1–2 cm)

Hind Print
Length: 0.7–1.3 in (1.8–3.3 cm)
Width: 0.5–0.9 in (1.3–2.3 cm)

Straddle
2–3.2 in (5–8 cm)

Stride
Bounding: 7–15 in (18–38 cm)

Size
Length with tail: 7–10 in (18–25 cm)

Weight
2.5–5 oz (70–140 g)

bounding

EASTERN CHIPMUNK

Tamias striatus

This large chipmunk is found in a variety of habitats, from the dense forest floor to open areas near buildings. Look for this delightful character throughout New England. You are more likely to see or hear this rodent, which is highly active during summer, than to notice its tracks. This chipmunk is happiest on the ground, but it will gladly climb sturdy oak trees to harvest juicy, ripe acorns. Eastern Chipmunks enter a deep sleep in winter, waking up from time to time to have a meal.

Chipmunks are so light that their tracks rarely show fine details. The forefeet each have four toes and the hind feet five. Chipmunks run on their toes, so the two heel pads of the forefeet seldom register; the hind feet have no heel pads. Their erratic track patterns show the hind feet registered in front of the forefeet, like many of their cousins. A chipmunk trail often leads to extensive burrows.

Similar Species: No other chipmunks live in New England. Midwinter tracks are more likely a squirrel's (pp. 74–79). Red Squirrel (p. 76) tracks are larger and have a larger straddle. Mouse (pp. 84–87) tracks are smaller.

Eastern Gray Squirrel

fore

hind

Fore Print
Length: 1–1.8 in (2.5–4.5 cm)
Width: 1 in (2.5 cm)

Hind Print
Length: 2.3–3 in (5.8–7.5 cm)
Width: 1.1–1.5 in (2.8–3.8 cm)

Straddle
3.8–6 in (9.5–15 cm)

Stride
Bounding: 0.7–3 ft (22–90 cm)

Size
Length with tail: 17–20 in (43–50 cm)

Weight
14–25 oz (400–710 g)

bounding

EASTERN GRAY SQUIRREL

Sciurus carolinensis

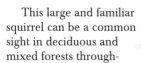

This large and familiar squirrel can be a common sight in deciduous and mixed forests through-out New England, even in urban areas. Active all year, the Eastern Gray Squirrel can leave a wealth of evidence, especially in winter as it scurries about digging up nuts that it buried during the previous fall.

The Eastern Gray Squirrel leaves a typical squirrel track when it runs or bounds. The hind prints fall slightly ahead of the fore prints. A clear fore print shows four toes with sharp claws, four fused palm pads and two heel pads. The hind print shows five toes and four palm pads and, if the full heel-length registers, it shows two small heel pads.

Similar Species: Red Squirrel (p. 76) prints are smaller. Eastern Chipmunks (p. 72) and flying squirrels (p. 78) make smaller tracks in a similar pattern, with narrower straddles. A rabbit or hare (pp. 60–63) makes a longer print pattern, and its forefeet rarely register side by side when it runs.

Red Squirrel

fore

hind

Fore Print
Length: 0.8–1.5 in (2–3.8 cm)
Width: 0.5–1 in (1.3–2.5 cm)

Hind Print
Length: 1.5–2.3 in (3.8–5.8 cm)
Width: 0.8–1.3 in (2–3.3 cm)

Straddle
3–4.5 in (7.5–11 cm)

Stride
Running: 8–30 in (20–75 cm)

Size
Length with tail:
    9–15 in (23–38 cm)

Weight
2–9 oz (55–260 g)

bounding

bounding
(in deep snow)

RED SQUIRREL
(Pine Squirrel, Chickaree)

Tamiasciurus hudsonicus

When you enter a
Red Squirrel's territory,
the inhabitant greets you
with a loud, chattering call.
Another obvious sign of this
forest dweller, which is found
throughout New England, is its large
middens—piles of cone scales and cores left beneath
trees—that indicate favorite feeding sites.

Active year-round in their small territories, Red Squirrels leave an abundance of trails that lead from tree to tree or down a burrow. These energetic animals mostly bound. This gait leaves groups of four prints, with the hind prints falling in front of the fore prints, which tend to be side by side (but not always). Four toes show on each fore print, and five show on each hind print. The heels often do not register when squirrels move quickly. In deeper snow the prints merge to form pairs of diamond-shaped tracks.

Similar Species: The larger Eastern Gray Squirrel (p. 74) is also found throughout New England. Chipmunks (p. 72) and flying squirrels (p. 78) make smaller tracks in a similar pattern, with a narrower straddle.

Northern Flying Squirrel

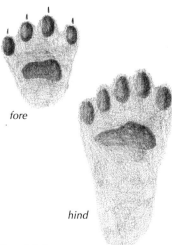

fore

hind

Fore Print
Length: 0.5–0.8 in (1.3–2 cm)
Width: 0.5 in (1.3 cm)

Hind Print
Length: 1.3–1.8 in (3.3–4.5 cm)
Width: 0.8 in (2 cm)

Straddle
3–3.8 in (7.5–9.5 cm)

Stride
Running: 11–30 in (28–75 cm)

Size
Length with tail: 9–12 in (23–30 cm)

Weight
4–6.5 oz (110–180 g)

sitzmark
into bounding

NORTHERN FLYING SQUIRREL

Glaucomys sabrinus

   This acrobat can be found in coniferous forests throughout all of New England. It prefers widely spaced forests, where it can glide from tree to tree by night, using the membranous flaps between its legs. Northern Flying Squirrels will den up together in a tree cavity for warmth.

   Because of its gliding, this squirrel does not leave as many tracks as other squirrels. Evidence is scarce in summer, but in winter you may come across a sitzmark—the distinctive pattern that an animal leaves when it lands in the snow—and a short bounding trail, which the squirrel makes as it rushes off to the nearest tree or to do some quick foraging. The bounding track pattern is typical of squirrels and other rodents, but with the hind feet registering only slightly in front of the forefeet—though often all four feet register in a row.

Similar Species: The Southern Flying Squirrel (*G. volans*) of southern New England is smaller and grayer. Red Squirrels (p. 76) usually make larger prints and rarely leave a sitzmark, but unclear tracks in deep snow can be identical. Chipmunks (p. 72) have smaller prints and a narrower straddle.

Norway Rat

fore

hind

Fore Print
Length: 0.7–0.8 in (1.8–2 cm)
Width: 0.5–0.7 in (1.3–1.8 cm)

Hind Print
Length: 1–1.3 in (2.5–3.3 cm)
Width: 0.8–1 in (2–2.5 cm)

Straddle
2–3 in (5–7.5 cm)

Stride
Walking: 1.5–3.5 in (3.8–9 cm)
Bounding: 9–20 in (23–50 cm)

Size
Length with tail:
    13–19 in (33–48 cm)

Weight
7–18 oz (200–510 g)

walking

NORWAY RAT
(Brown Rat)
Rattus norvegicus

Active both day and night, this despised rat is widespread almost anywhere that humans have decided to build homes. Not entirely dependent on people, it may live in the wild as well.

The fore print shows four toes, and the hind print shows five. When it bounds, this colonial rat leaves four-print groups, with the hind prints in front of the diagonally placed fore prints. Sometimes one of the hind feet direct registers on a fore print, creating a three-print group. More commonly, rats leave an alternating walking pattern, with the larger hind prints close to or overlapping the fore prints; the hind heel does not show. In snow, the tail often leaves a dragline. Rats live in groups, so you may find many trails together, often leading to their 5-inch (2-cm) wide burrows.

Similar Species: The Black Rat (*Rattus rattus*) makes comparable tracks and is found in most of New England. Mice (pp. 84–87) prints are much smaller. Red Squirrel (p. 76) tracks show distinctive squirrel characteristics.

Meadow Vole

fore

hind

Fore Print
Length: 0.3–0.5 in (0.8–1.3 cm)
Width: 0.3–0.5 in (0.8–1.3 cm)

Hind Print
Length: 0.3–0.6 in (0.8–1.5 cm)
Width: 0.3–0.6 in (0.8–1.5 cm)

Straddle
1.3–2 in (3.3–5 cm)

Stride
Walking/Trotting:
   1.3–3 in (3.3–7.5 cm)
Bounding: 4–8 in (10–20 cm)

Size
Length with tail:
   5.5–8 in (14–20 cm)

Weight
0.5–2.5 oz (14–70 g)

walking

*bounding
(in snow)*

MEADOW VOLE
(Field Mouse)
Microtus pennsylvanicus

With so many vole species in the region, positive trail identification is next to impossible. Note, though, that the Meadow Vole frequents damp or wet habitats.

When clear (which is seldom), vole fore prints show four toes, and hind prints show five. A vole's walk and trot both leave a paired alternating track pattern with a hind print occasionally direct registered on a fore print. Voles usually opt for a faster bounding; the resulting print pairs show the hind prints registered on the fore prints. These voles lope quickly across open areas, creating a three-print track pattern. In winter, voles stay under the snow; when it melts, look for distinctive piles of cut grass from their ground nests. The bark at the bases of shrubs may show tiny teeth marks left by gnawing. In summer, well-used vole paths appear as little runways in the grass.

Similar Species: The Woodland Vole (*M. pinetorum*), the Southern Red-backed Vole (*Cletherionomys gapperi*) and the less-common Rock Vole (*M. chrotorrhinus*) also inhabit the region. The Deer Mouse's (p. 84) track pattern has shorter strides, and its fore prints show only four toes.

Deer Mouse

bounding group

Fore Print
Length: 0.3–0.4 in (0.8–1 cm)
Width: 0.3–0.4 in (0.8–1 cm)

Hind Print
Length: 0.3–0.5 in (0.8–1.3 cm)
Width: 0.3–0.4 in (0.8–1 cm)

Straddle
1.4–1.8 in (3.6–4.5 cm)

Stride
Bounding: 5–12 in (15–30 cm)

Size
Length with tail:
    6–12 in (15–30 cm)

Weight
0.5–1.3oz (14–35 g)

bounding

bounding (in snow)

DEER MOUSE

Peromyscus maniculatus

The highly adaptable Deer Mouse—the most abundant mammal in this region—is found anywhere from arid valleys all the way up to alpine meadows. It is seldom seen because it is nocturnal. The Deer Mouse may enter buildings in winter, where it will stay active.

The fore prints each show four toes, three palm pads and two heel pads. The hind prints show five toes and three palm pads; the heel pads rarely register. It takes perfect, soft mud to get clear prints from such a tiny mammal. Bounding tracks, most noticeable in snow, show the hind prints falling in front of the fore prints. In soft snow the prints may merge to look like larger pairs of prints, with tail drag evident. A trail may lead up a tree or down into a burrow.

Similar Species: Tracks of many less common species of mice are identical. House Mouse (*Mus musculus*) tracks are very similar, but they are associated more with humans. Voles (p. 82) tend to trot and have a much shorter bounding track pattern. Jumping mouse (p. 86) prints may be similar in size, but they show long, thin toes. Chipmunks (p. 72) have a wider straddle. Shrews (p. 88) have a narrower straddle.

Meadow Jumping Mouse

*bounding
group*

Fore Print
Length: 0.3–0.5 in (0.8–1.3 cm)
Width: 0.3–0.5 in (0.8–1.3 cm)
Hind Print
Length: 0.5–1.3 in (1.3–3.3 cm)
Width: 0.5–0.7 in (1.3–1.8 cm)
Straddle
1.8–1.9 in (4.5–4.8 cm)
Stride
Bounding: 7–18 in (18–45 cm)
In alarm: 3–6 ft (90–180 cm)
Size
Length with tail: 7–9 in (18–23 cm)
Weight
0.6–1.3 oz (17–35 g)

bounding

MEADOW JUMPING MOUSE
Zapus hudsonius

Though the Meadow Jumping Mouse lives throughout New England, it has a long, deep winter hibernation (about six months!) when no tracks are found. During the warmer months, in suitable substrate near grassy areas, their tracks may be abundant.

Jumping mouse tracks are distinctive if you do find them. The two smaller forefeet register between the long hind feet; the long heels do not always register and some prints show just the three long middle toes. The toes on the forefeet may splay so much that the side toes point backward. When bounding, these mice make short leaps. The tail may leave a dragline in soft mud or unseasonable snow. Clusters of cut grass stems, about 5 inches (13 cm) long, lying in meadows, are a more abundant sign of this rodent.

Similar Species: The Woodland Jumping Mouse (*Napaeozapus insignis*) is found in wet, wooded areas. Deer Mouse (p. 84) tracks may have the same straddle. Heel-less hind prints may be mistaken for a vole's (p. 82)—or a small bird's (p. 110) or an amphibian's (pp. 112–117).

Masked Shrew

Fore Print
Length: 0.2 in (0.5 cm)
Width: 0.2 in (0.5 cm)

Hind Print
Length: 0.6 in (1.5 cm)
Width: 0.3 in (0.8 cm)

Straddle
0.8–1.3 in (2–3.3 cm)

Stride
Bounding: 1.2–3 in (3–7.5 cm)

Size
Length with tail:
2.3–4.5 in (7–11 cm)

Weight
0.1–0.3 oz (3–9 g)

bounding

MASKED SHREW
Sorex cinereus

Though several species of tiny, frenetic shrews are found in New England, if you find tracks, a likely candidate is the widespread and adaptable Masked Shrew. This small shrew prefers moist fields, marshes, bogs or woodlands, but it can also be found in higher and drier grasslands. Its rapid activity makes it difficult to observe closely.

In its energetic and unending quest for food, a shrew usually leaves a four-print bounding pattern, but it may leave an alternating walking pattern. The individual prints in a group are often indistinct, but in mud or shallow, wet snow, you can even count the five toes on each print. In deeper snow, a shrew's tail often leaves a dragline. If a shrew tunnels under the snow, it may leave a snow ridge on the surface. A shrew's trail may disappear down a burrow.

Similar Species: The Smoky Shrew (*S. fumeus*) and Pygmy Shrew (*S. hoyi*), which are more common in moist, wooded areas, leave similar tracks. The Water Shrew (*S. palustris*), often found by cold mountain streams, and the widespread, abundant Northern Short-tailed Shrew (*Blarina brevicauda*) are larger. Mouse (pp. 84–87) fore prints show only four toes.

a molehill of the Star-nosed Mole

some molehills and ridges of the Eastern Mole

STAR-NOSED MOLE

Condylura cristata

This peculiar character is found throughout most of New England. The Star-nosed Mole is easily identified by the strange tentacle-like protrusions on its nose, thought to help it find food in its dark and often subterranean world. It spends more time out of its burrow than other moles do.

Typical signs of this mole include the big piles of mud pushed out of its burrows. Because of its preference for swimming and wet areas, look for this mole's hills along the banks of streams and rivers and in raised areas around marshes and wet fields. Clear prints from its long-clawed feet will rarely be found. The Star-nosed Mole remains under the snow during winter and swims under the ice, so winter tracks are seldom seen.

Similar Species: The Eastern Mole (*Scalopus aquaticus*), in southern New England, leaves the familiar distinctive molehills that are often connected by raised ridges created by digging tunnels near the surface (as illustrated at left).

BIRDS, AMPHIBIANS & REPTILES

A guide to the animal tracks of New England is not complete without some consideration of the birds and amphibians found in the region.

Several bird species have been chosen to represent the main types common to this region. Remember, however, that individual bird species are not easily identified by track alone. Bird tracks can often be found in abundance in snow and are clearest in shallow, wet snow. The shores of streams and lakes are very reliable locations in which to find bird tracks—the mud there can hold a clear print for a long time. The sheer number of tracks made by shorebirds and waterfowl can be astonishing. Though some bird species prefer to perch in trees or soar across the sky, it can be entertaining to follow the tracks of birds that spend a lot of time on the ground. They can spin around in circles and lead you in all directions. The trail may suddenly end as the bird takes flight, or it might terminate in a pile of feathers, the bird having fallen victim to a hungry predator.

Many amphibians and turtles depend on moist environments, so look in the soft mud along the shores of lakes and ponds for their distinctive tracks. Though you may be able to distinguish frog tracks from toad tracks, because these amphibians generally move differently, it can be very difficult to identify the species. In drier environments, reptiles, which thrive in dryness, outnumber the amphibians. Unfortunately, because of this preference for dry terrain, most reptiles seldom leave good tracks for us to identify.

Mallard

Print
Length: 2–2.5 in (5–6.5 cm)
Straddle
4 in (10 cm)
Stride
to 4 in (10 cm)
Size
23 in (58 cm)

MALLARD

Anas platyrhynchos

female

male

This dabbling duck—the male a familiar sight with its striking green head—is common in open areas near lakes and ponds. Its webbed feet leave prints that can often be seen in abundance along the muddy shores of just about any waterbody, including those in urban parks.

The webbed foot of the Mallard has three long toes that all point forward. Though the toes register well, the webbing between the toes does not always show in the print. The inward-pointing feet give the Mallard a pigeon-toed appearance, which perhaps accounts for its waddling gait, a characteristic for which ducks are known.

Similar Species: The many gulls (p. 96) and other dabbling ducks have similar prints. Exceptionally large prints are from geese or swans (*Cygnus* spp.).

Herring Gull

Print
Length: 3.5 in (9 cm)
Straddle
4–6 in (10–15 cm)
Stride
4.5 in (11 cm)
Size
Length: 23–25 in (58–65 cm)

HERRING GULL

Larus argentatus

Herring gulls, with their long wings and webbed toes, are strong long-distance fliers as well as excellent swimmers. They are found throughout coastal New England and are concentrated in great numbers on the coast.

Gulls leave slightly asymmetrical tracks that show three toes. They have claws that register outside the webbing, and the clawmarks are usually attached to the footprint. Most gulls have quite a swagger to their gait, and they leave a trail with the tracks turned strongly inward.

Similar Species: Gull species cannot be reliably identified by track alone, but smaller species have conspicuously smaller tracks. Duck tracks (p. 94) are difficult to distinguish from gull tracks.

Great Blue Heron

Print
Length: to 6.5 in (17 cm)
Straddle
8 in (20 cm)
Stride
9 in (23 cm)
Size
4.2–4.5 ft (1.3–1.4 m)

GREAT BLUE HERON

Ardea herodias

The refined and graceful image of this large heron symbolizes the precious wetlands in which it patiently hunts for food. Usually still and statuesque as it waits for a meal to swim by, this heron will have cause to walk from time to time, perhaps to find a better hunting location. Look for its large, slender tracks along the banks or mud-flats of waterbodies.

Not surprisingly, a bird that lives and hunts with such precision walks in a similar fashion, leaving straight tracks that fall in a nearly straight line. Look for the slender rear toe in the print.

Similar Species: Other herons make similar tracks that vary in size according to the size of the bird. Cranes (*Grus* spp.) have similar, possibly larger, tracks, but their smaller hind toes do not register.

Common Snipe

Print
Length: 1.5 in (3.8 cm)
Straddle
to 1.8 in (4.5 cm)
Stride
to 1.3 in (3.3 cm)
Size
11–12 in (28–30 cm)

COMMON SNIPE
Gallinago gallinago

This short-legged character is a resident of marshes and bogs, where its neat prints can often be seen in mud. Snipes are quite secretive when on the ground, and so you may be surprised if one suddenly flushes out from beneath your feet. One good sign that snipes are in the area is their eerie whistle as they dive from the sky.

The Common Snipe's neat prints show four toes, including a small rear toe that points inward. The bird's short legs and stocky body give it a very short stride.

Similar Species: Many shorebirds, including the Spotted Sandpiper (p. 102), leave similar tracks. American Woodcock (*Scolopax minor*) tracks tend to be a little larger: woodcocks are widespread in the northeast and in forests and open areas.

Spotted Sandpiper

Print
Length: 0.8–1.3 in (2–3.3 cm)
Straddle
to 1.5 in (3.8 cm)
Stride
Erratic
Size
7–8 in (18–20 cm)

SPOTTED SANDPIPER

Actitis macularia

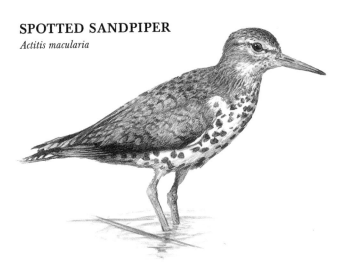

The bobbing tail of the Spotted Sandpiper is a common sight on the shores of lakes, rivers and streams, but you will usually find just one of these territorial birds in any given location. Because of its excellent camouflage, likely the first you will see of this bird will be when it flies away, its fluttering wings close to the surface of the water.

As this sandpiper teeters up and down on the shore, it leaves trails of three-toed prints. Its fourth toe is very small and faces off to one side at an angle. Sandpiper tracks can have an erratic stride.

Similar Species: All sandpipers and plovers, including the Killdeer (*Charadrius vociferus*), leave similar tracks, although there is much diversity in size. The Common Snipe (p. 100) makes larger tracks.

Ruffed Grouse

Print
Length: 2–3 in (5–7.5 cm)
Straddle
2–3 in (5–7.5 cm)
Stride
Walking: 3–6 in (7.5–15 cm)
Size
15–19 in (34–48 cm)

RUFFED GROUSE

Bonasa umbellus

This ground-dweller prefers the quiet seclusion of coniferous forests in winter, so that will be the best place to find its tracks. If you follow them quietly, you may be startled when the grouse bursts from cover underneath your feet. Its excellent camouflage usually affords it good protection.

The three thick front toes leave very clear impressions, but the short rear toe, which is angled off to one side, does not always show up so well. This bird's neat, straight trail appears to reflect its cautious approach to life on the forest floor.

Similar Species: The Spruce Grouse (*Dendragapus canadensis*) leaves similar tracks, but they may be enlarged and obscured by the winter feathers that this bird grows on its feet.

Great Horned Owl

Strike
Width: to 3 ft (90 cm)
Size
22 in (55 cm)

GREAT HORNED OWL

Bubo virginianus

Often seen resting quietly in trees by day, this wide-ranging owl prefers to hunt at night. An accomplished hunter in snow, the owl strikes through the snow with its talons, leaving an untidy hole that may be surrounded by wing and tail-feather imprints. If it registers well, this 'strike' can be quite a sight. The feather imprints are made as the owl struggles to take off with possibly heavy prey. An ungraceful walker, it prefers to fly away from the scene.

You may stumble across a strike and guess that the owl's target could have been a vole (p. 82) scurrying around underneath the snow. Or you may be following the surface trail of an animal to find that it abruptly ends with this strike mark, where the animal has been seized by an owl or perhaps a hawk or raven.

Similar Species: Other large, predatory birds also leave strike marks. If the prey left no approaching trail, the strike mark is likely an owl's, because owls hunt by sound. If there is a trail, the strike mark, usually with much sharper feather imprints, could be from a hawk or a Common Raven (*Corvus corax*), both of whom hunt by sight.

American Crow

Print
Length: 2.5–3 in (6.5–7.5 cm)
Straddle
1.5–3 in (3.8–7.5 cm)
Stride
Walking: 4 in (10 cm)
Size
16 in (40 cm)

AMERICAN CROW

Corvus brachyrhyncos

   The black silhouette of the American Crow can be a common sight in a variety of habitats. A crow will frequently come down to the ground and contentedly strut around with a confidence that hints at its intelligence. Its loud *caw* can be heard from quite a distance; crows can be especially noisy when they are mobbing an owl or hawk.

   The American Crow typically leaves an alternating walking track pattern. Its prints show three sturdy toes pointing forward and one toe pointing backward. When a crow is in need of greater speed, perhaps for takeoff, it bounds along, leaving irregular pairs of diagonally placed prints with a longer stride between each pair.

Similar Species: Other corvids also spend a lot of time on the ground and make similar tracks—in coastal areas, one example is the Fish Crow (*C. ossifragus*). Jays also make similar tracks. The much larger Common Raven (*C. corax*) leaves tracks up to 4 inches (10 cm) long, with a stride of 6 inches (15 cm).

Dark-eyed Junco

Print
Length: to 1.5 in (3.8 cm)
Straddle
1–1.5 in (2.5–3.8 cm)
Stride
Hopping: 1.5–5 in (3.8–13 cm)
Size
5.5–6.5 in (14–17 cm)

DARK-EYED JUNCO
Junco hyemalis

This common small bird typifies the many small hopping birds found in the region. Each foot has three forward-pointing toes and one longer toe at the rear. The best prints are left in snow, although in deep snow the toe detail is lost; the footprints may show some dragging between the hops.

A good place to study this type of prints is near a birdfeeder. Watch the birds scurry around as they pick up fallen seeds, then have a look at the prints left behind. For example, juncos are attracted to seeds that chickadees scatter as they forage for sunflower seeds in the birdfeeder. Also look for tracks under coniferous trees, where juncos feed on fallen seeds in winter.

Similar Species: Toe size may help with identification—larger birds make larger prints. In powdery snow, junco tracks could be mistaken for mouse (pp. 84–87) tracks—follow the trail to see if it disappears down a hole or into thin air.

Frogs

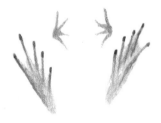

Straddle
to 3 in (7.5 cm)

FROGS

Wood Frog

The best place to look for frog tracks is along the muddy fringes of waterbodies. A frog's hopping action results in its two small forefeet registering in front of its long-toed hind prints. Frog tracks vary greatly in size, depending on species and age.

The smallest frogs include the treefrogs, such as the Spring Peeper (*Pseudoacris crucifer*). Its size—a length of just 1.5 inches (3.8 cm)—and its preference for thick undergrowth and shrubs near the water make its tracks a rare sight. The larger Gray Treefrog (*Hyla chrysoscelis* and *H. versicolor*) spends most of its time in trees, coming down to breed and sing at night. The widespread Wood Frog (*Rana sylvatica*) is larger, at about 3.5 inches (9 cm) long; it can often be found in dry woodland areas, seemingly far from water. The Green Frog (*R. clamitans*) and Pickerel Frog (*R. palustris*) are also widespread. Unusually large tracks are surely from the robust Bullfrog (*R. catesbeiana*). Growing to 8 inches (20 cm) in length, it is North America's largest frog.

Toads

hind *fore*

Straddle
to 2.5 in (6.5 cm)

TOADS

Woodhouse's Toad

There are fewer toads species than frog species in New England. The best place to look for the tracks of toads is, as with frogs, along the muddy fringes of waterbodies, but they can occasionally be found in drier areas, for example, as unclear trails in dusty patches of soil.

The toad most likely to be encountered, and the most widespread, is the American Toad (*Bufo americanus*), which lives in many different moist habitats. Woodhouse's Toad (*B. woodhousei*) is found scattered throughout the region in temporary pools and ditches. Toads in New England can be up to 4.5 inches (11 cm) in length.

In general, toads walk and frogs (p. 112) hop, but toads are pretty capable hoppers, too, especially when being hassled by overly enthusiastic naturalists. Toads leave rather abstract prints as they walk. The heels of the hind feet do not register. On less firm surfaces, the toes often leave draglines.

Salamanders & Newts

Straddle
to 7.5 cm (3 in)

salamander walking

SALAMANDERS & NEWTS

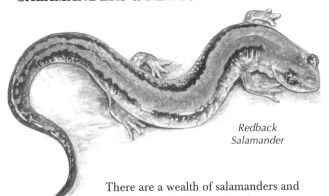

*Redback
Salamander*

There are a wealth of salamanders and newts in New England's moist and wet areas. Among the more abundant and widespread of these long, slender, lizard-like amphibians is the Eastern Newt (*Notophthalmus viridescens*), which can grow up to 5.5 inches (14 cm) long. After a fresh rain, Eastern Newts emerging from ponds may leave small trails in the mud.

There are more species of salamanders than newts in New England. The Redback Salamander (*Plethodon cinereus*), which can reach 5 inches (13 cm) in length, lives throughout the region in mixed forests and some coniferous forests. The stout-bodied Spotted Salamander (*Ambystoma maculatum*), which can reach up to 10 inches (25 cm) in length, lives around pools and flooded areas in hardwood forests.

In general, a salamander or newt fore print shows four toes, and the larger hind print shows five. However, print detail is often obscured by the animal's dragging belly or by the swinging of its thick tail across the trail.

Skinks

walking

SKINKS

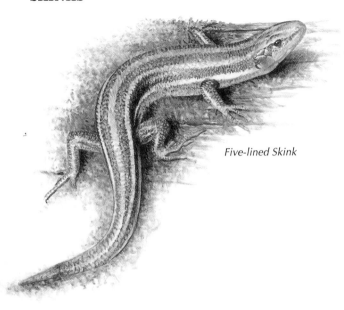

Five-lined Skink

   Skinks, which are reptiles and not amphibians, leave tracks similar to those of a salamander, but their toes are longer and more slender. If you find skink tracks, the most likely candidate is the Five-lined Skink (*Eumeces fasciatus*), which can grow up to 8 inches (20 cm) in length. It favors moist woodlands in the southern parts of New England.

Turtles

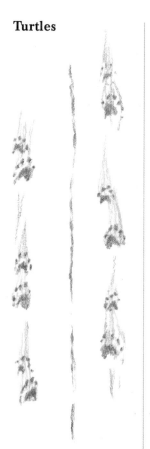

typical turtle walking

Snapping Turtle walking

TURTLES

Wood Turtle

Turtles, those ancient denizens of the water world, will happily slip into the murky depths of the water to avoid detection. They do, however, come out from time to time to feed or bask in the sunshine. Look for their distinctive tracks alongside ponds, rivers and moist areas. Note that some turtles, such as the huge Snapping Turtle (*Chelydra serpentina*), prefer to stay in the water and rarely come out.

One of the most widespread and prettiest turtles, often seen basking, is the Painted Turtle (*Chrysemys picta*); it can reach almost 10 inches (25 cm) in length. Slightly smaller is the Wood Turtle (*Clemmys insculpta*), which likes moist forested areas and woodland streams. The Spotted Turtle (*Clemmys guttata*), which reaches only 5 inches (13 cm) in length, prefers boggy areas.

With its large shell and short legs, a turtle leaves a track that is wide relative to the length of its stride—its straddle is about half its body length. Although longer-legged turtles can raise their shells off the ground, short-legged species may let their shells drag, as shown in their tracks. The tail may leave a straight dragline in the mud. On firmer surfaces, look for distinct claw marks.

Snakes

typical snake track

SNAKES

Common Garter Snake

There are many snakes to be found throughout New England, with much greater diversity in the warmer south. Snakes are all long and slender—their tracks appear so similar that identification among the species is next to impossible. In fact, because a snake lacks feet and leaves a track that is just a gentle meander, it is very challenging even to establish in which direction the snake is traveling.

The most widespread and frequently encountered snake is the Common Garter Snake (*Thamnophis sirtalis*). Found throughout the region, often close to wet or moist areas, this harmless snake can reach 4.3 feet (1.3 m) in length.

Other widespread species include the Red-bellied Snake (*Storeria occipitomaculata*), which prefers hilly woodlands and reaches a length of up to 16 inches (40 cm). Only slightly larger, the Smooth Green Snake (*Opheodrys vernalis*) inhabits grassy meadows and fields along forest edges. In sandy, open areas of the south, a snake track might be from the Eastern Hognose Snake (*Heterodon platirhinos*).

TRACK PATTERNS & PRINTS

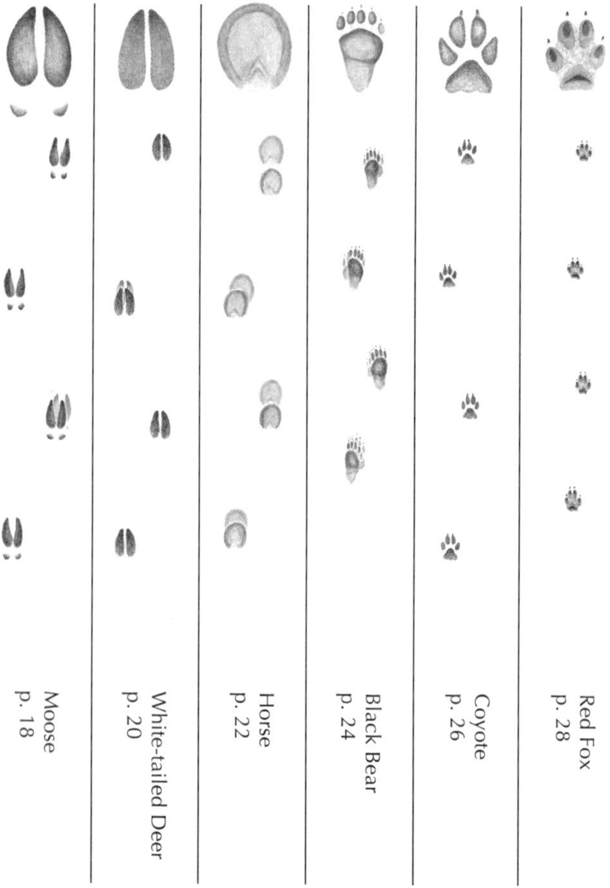

Moose p. 18	White-tailed Deer p. 20	Horse p. 22	Black Bear p. 24	Coyote p. 26	Red Fox p. 28

Gray Fox
p. 30

Mountain Lion
p. 32

Lynx
p. 34

Bobcat
p. 36

Domestic Cat
p. 38

Raccoon
p. 40

TRACK PATTERNS & PRINTS

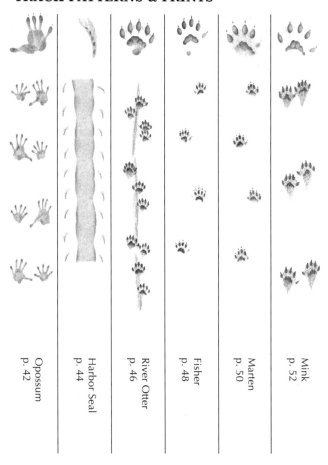

Opossum
p. 42

Harbor Seal
p. 44

River Otter
p. 46

Fisher
p. 48

Marten
p. 50

Mink
p. 52

126

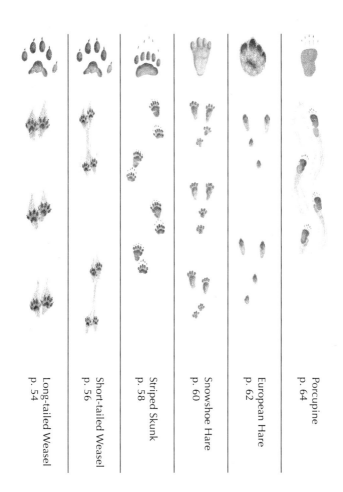

TRACK PATTERNS & PRINTS

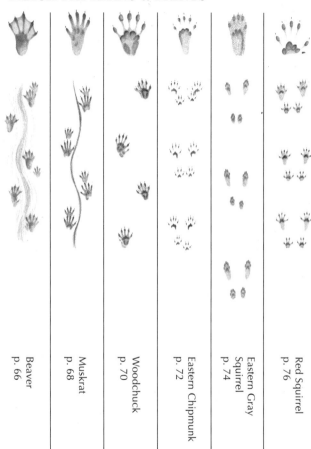

Beaver
p. 66

Muskrat
p. 68

Woodchuck
p. 70

Eastern Chipmunk
p. 72

Eastern Gray
Squirrel
p. 74

Red Squirrel
p. 76

Masked Shrew
p. 88

Meadow
Jumping Mouse
p. 86

Deer Mouse
p. 84

Meadow Vole
p. 82

Norway Rat
p. 80

Northern
Flying Squirrel
p. 78

129

TRACK PATTERNS & PRINTS

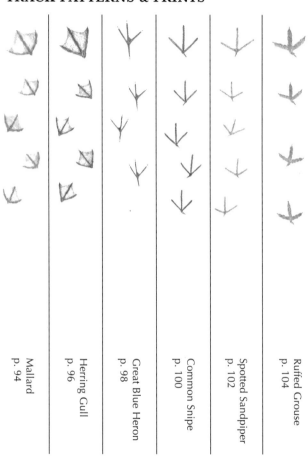

Mallard
p. 94

Herring Gull
p. 96

Great Blue Heron
p. 98

Common Snipe
p. 100

Spotted Sandpiper
p. 102

Ruffed Grouse
p. 104

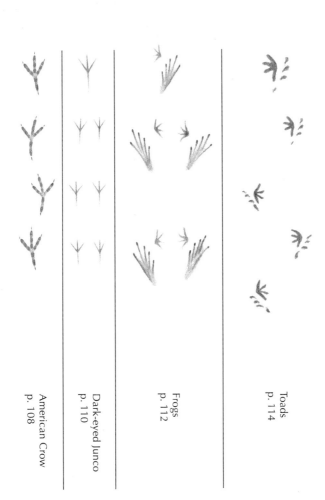

Toads
p. 114

Frogs
p. 112

Dark-eyed Junco
p. 110

American Crow
p. 108

TRACK PATTERNS & PRINTS

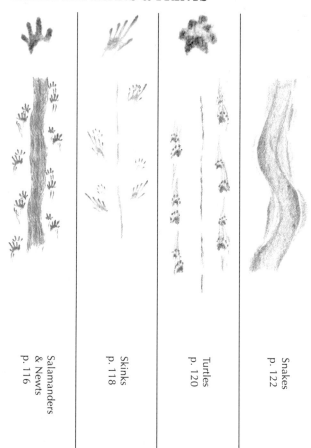

Salamanders
& Newts
p. 116

Skinks
p. 118

Turtles
p. 120

Snakes
p. 122

HOOFED PRINTS

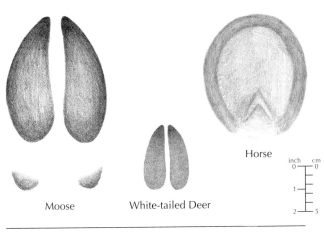

Moose

White-tailed Deer

Horse

HIND PRINTS

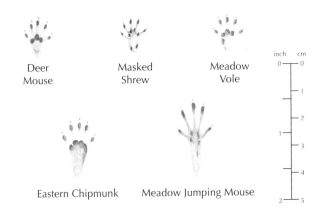

Deer
Mouse

Masked
Shrew

Meadow
Vole

Eastern Chipmunk

Meadow Jumping Mouse

HIND PRINTS

Norway
Rat

Northern
Flying Squirrel

Red
Squirrel

Woodchuck

Eastern
Gray Squirrel

Muskrat

inch   cm
0 — 0

1 —

2 — 5

Porcupine

Raccoon

Opossum

134

HIND PRINTS

Snowshoe
Hare

European
Hare

Beaver

Black Bear

inch cm
0   0

1

2   5

 135

FORE PRINTS

Short-tailed Weasel

Long-tailed Weasel

Mink

Striped Skunk

Domestic Cat

Gray Fox

Red Fox

Coyote

Bobcat

Marten

River Otter

Fisher

inch cm
0 — 0

1 —

2 — 5

Mountain Lion

Lynx

BIBLIOGRAPHY

Behler, J.L., and F.W. King. 1979. *Field Guide to North American Reptiles and Amphibians.* National Audubon Society. New York: Alfred A. Knopf.

Brown, Roy, John Ferguson, Michael Lawrence and David Lees. 1987. *Tracks and Signs of the Birds of Britain and Europe. An Identification Guide.* London: Christopher Helm.

Burt, W.H. 1976. *A Field Guide to the Mammals.* Boston: Houghton Mifflin Company.

Farrand, J., Jr. 1995. *Familiar Animal Tracks of North America.* National Audubon Society Pocket Guide. New York: Alfred A. Knopf.

Forrest, L.R. 1988. *Field Guide to Tracking Animals in Snow.* Harrisburg: Stackpole Books.

Halfpenny, J. 1986. *A Field Guide to Mammal Tracking in North America.* Boulder: Johnson Publishing Company.

Headstrom, R. 1971. *Identifying Animal Tracks.* Toronto: General Publishing Company.

Murie, O.J. 1974. *A Field Guide to Animal Tracks.* The Peterson Field Guide Series. Boston: Houghton Mifflin Company.

Rezendes, P. 1992. *Tracking and the Art of Seeing: How to Read Animal Tracks and Signs.* Vermont: Camden House Publishing.

Stall, C. 1989. *Animal Tracks of the Rocky Mountains.* Seattle: The Mountaineers.

Stokes, D., and L. Stokes. 1986. *A Guide to Animal Tracking and Behaviour.* Toronto: Little, Brown and Company.

Wassink, J.L. 1993. *Mammals of the Central Rockies.* Missoula: Mountain Press Publishing Company.

Whitaker, J.O., Jr. 1996. *National Audubon Society Field Guide to North American Mammals.* New York: Alfred A. Knopf.

INDEX

Page numbers in **boldface** type refer to the primary (illustrated) treatments of animal species and their tracks.

ABOUT THE AUTHORS

Ian Sheldon, an accomplished artist, naturalist and educator, has lived in South Africa, Singapore, Britain and Canada. Caught collecting caterpillars at the age of three, he has been exposed to the beauty and diversity of nature ever since. He was educated at Cambridge University and the University of Alberta. When he is not in the tropics working on conservation projects or immersing himself in our beautiful wilderness, he is sharing his love for nature. Ian enjoys communicating this passion through the visual arts and the written word.

Tamara Hartson, equipped from the age of six with a canoe, a dip net and a note pad, grew up with a fascination for nature and the diversity of life. She has a degree in environmental conservation sciences and has photographed and written about the biodiversity in Bermuda, the Galapagos Islands, the Amazon Basin, China, Tibet, Vietnam, Thailand and Malaysia.

Mark Elbroch resides in southern New Hampshire. Teaching courses throughout New England, he spreads passion for tracking, birds and everything wild. Mark has also contributed to tracking research projects across the country, including Montana, Washington and North Carolina. It is his hope—as people learn more about tracking—that we will better preserve wild places and the wildlife that inhabits them.

LONE
PINE

Look for these other great Lone Pine titles:

Mushrooms of Northeast North America, Midwest to New England
By George Barron
ISBN 1-55105-201-6 • 336 pages • Over 600 color photographs
$19.95

A full-color photographic field guide to more than 600 species of mushrooms and fungi of the northern United States, from the Midwest to New England. Spectacular photographs and excellent species information combine to make this a must-have reference.

Weeds of the Northern U.S. and Canada
By Richard Dickinson & France Royer
ISBN 1-55105-221-0 • 400 pages • 800 color photographs
$21.95

This impressive, richly illustrated field guide identifies more than 150 noxious weeds. Color photographs show the weeds at the five critical stages and help to distinguish another 100 related species. Information on weed legislation by state is also listed. A first of its kind, this book will be an extraordinary resource for a multitude of users, whether farmer, landscaper, weed specialist or gardener.

Birds of Boston
By Chris C. Fisher & Andy Bezener
ISBN 1-55105-182-6 • 160 pages • color illustrations
$9.95

This easy-to-use and beautifully illustrated guidebook will help you identify the feathered strangers nibbling at the feeder in your backyard or singing from a nearby tree. It's packed with notes on 127 common and interesting bird species in and around Boston and Massachusetts Bay.

US Orders
1-800-518-3541 Phone
1-800-548-1169 Fax